SPACE

Could life exist beyond Earth?

Caroline Smith, Natasha Almeida,
Ashley King, Peter Grindrod
and Sara Russell

Published by the Natural History Museum, London

First published by the Natural History Museum,
Cromwell Road, London SW7 5BD
© The Trustees of the Natural History Museum,
London, 2025

The Authors have asserted their rights to be identified as
the Authors of this work under the Copyright, Designs
and Patents Act 1988

ISBN 9780565095697

A catalogue record for this book is available
from the British Library
10 9 8 7 6 5 4 3 2 1

Designed by Bobby Birchall, Bobby&Co
Reproduction by Saxon Digital Services
Printed by Toppan Leefung Printing Limited, China
This book is printed using 100% mineral free oil inks

Front cover: Galaxies: © overlays-textures/shutterstock;
dust storm: © Hussain Warraich/shutterstock; blue ice:
© Katvic/shutterstock; Europa: © Artsiom/shutterstock;
Mars: © Williams Harking/shutterstock

Contents

Introduction

Caroline Smith

Humans have long looked to the sky and pondered on one of the biggest philosophical questions 'are we alone?' With the invention of the telescope in the early 17th century, famous astronomers like Galileo, Kepler and Huygens trained their eyes on celestial objects like our Moon, Mars, Saturn and Jupiter. In the early 19th century, advances in telescope design and optics allowed for the discovery of smaller objects like asteroids and more detailed observations of the surfaces of planets in the later 19th and early 20th centuries. Many of the investigations carried out by these early astronomers were looking for signs of life in the cosmos.

At around the same time as the first asteroids were discovered, scientists studying strange rocks that had been seen to fall to Earth were beginning to realize that these objects had textures and chemical compositions that were nothing like rocks found on Earth. They shared common characteristics between them – they were covered in a black crust and contained iron-nickel metal. It was initially thought that these rocks, which we now know as meteorites, might have been erupted from lunar volcanoes. However, this theory was dismissed with the realization that the Moon did not show evidence of active volcanism and that, as more meteorites were analyzed in detail, they showed a great diversity in composition and so could not all be from the same place. During the 19th century, increasing numbers of asteroids were being discovered orbiting our Sun between the orbits of Mars and Jupiter. Scientists suggested that these small bodies might be the remnants of an exploded planet (we now know this is incorrect), and so it was not unreasonable to link the meteorites falling on Earth to these asteroids in space (this is correct!). Meteorites are now recognized as rare and precious samples of bodies within our Solar System and as originating from different types of asteroids, the Moon and the planet Mars. Studying meteorites in detail has provided a wealth of information about the formation of our Solar System and its evolution. Some meteorites contain water and organic molecules, both critical ingredients needed for life. We can see evidence in some Martian meteorites that the minerals have been affected by water, showing that in the past liquid water was present on Mars – and liquid water is an important factor to sustain life.

With the dawn of the space age in the late 1950s, scientists and engineers once more turned their eyes to the sky, and the next phase of human exploration began. Within 30 years of the first satellite launch (Sputnik 1 in 1957) space probes were launched to visit our Moon, land on Venus and Mars, and visit the outer planets and their moons. Data from these early missions provided tantalizing evidence that other places in our Solar System had environments that were habitable and conducive to life, either in the past or now. These discoveries laid the foundation for more complex missions to Mars and the moons of Jupiter and Saturn, places we might have the best chances of finding signs of fossilized life or perhaps even life existing today. Today, NASA has two rovers actively working on Mars, studying Mars geology and its ancient environment. The Perseverance rover is also selecting rock samples, which scientists hope will be returned to Earth in the 2030s, and which can be studied for the telltale physical and chemical fingerprints left behind by ancient life. Two missions have been launched to study three of Jupiter's moons – ESA's Juice mission and NASA's Europa Clipper. These missions are tasked with investigating the moons' sub-surface oceans and looking for the physical and chemical signs that these oceans could support life. ESA is also actively studying Mars and the ambitious Rosalind Franklin rover is scheduled to be launched in 2028. This rover has instruments specifically designed to detect the presence of sub-surface life on Mars today. In the late 2020s, NASA plans to launch a mission called Dragonfly to Saturn's moon Titan, another moon that has been identified as having a potentially habitable environment and which scientists use as a model for what our own Earth was like early in its history.

Humankind's exploration of space started with telescope observations in the 17th century, and we are now using advanced telescopes to search outside our Solar System for planets elsewhere in our galaxy that may have the right conditions for life. Planets that orbit stars outside our Solar System are known as exoplanets. Since the first exoplanets were discovered in 1992, there are about about 5,800 confirmed exoplanets in approximately 4,300 planetary systems (as of November 2024). Astronomers are now using the James Webb Space Telescope – the largest, most complex and most powerful telescope ever launched into space – to peer deep into our galaxy and examine these exoplanets in detail. Instruments on the telescope can measure the mass of the exoplanet, analyze the temperature and composition of the atmospheres, and take direct images.

For the first time in our species' history we are on the cusp of making discoveries that will be able to provide evidence to allow us to answer the fundamental question of 'are we alone?' Is our Solar System unusual? Why has the Solar System been successful in creating a planet that supports life (our Earth)? Did life exist, or does it exist now in other places in our Solar System? Are there other planetary systems in our galaxy that have the right conditions for life? If life exists elsewhere, would it look similar to life on Earth or vastly different? Scientists increasingly think that some form of life does exist elsewhere in space..... What do you think?

Earth

Natasha Almeida

The Sun, our star, was formed by the collapse of a dense cloud of dust and gas called the solar nebula. The remaining material that escaped being incorporated into the Sun, perhaps around 1% of this original cloud, was left spinning around the star in what scientists refer to as the protoplanetary disk. This disk, made up of dust grains and gas, formed every other planetary body in the Solar System. The formation of the Sun and the beginning of our Solar System is called time zero. To give a date to this time zero, scientists have investigated the earliest solid grains – calcium-aluminium-rich inclusions (CAIs) – that are found in the most primitive of meteorites, such as the carbonaceous chondrites, meteorites that have not been changed by extensive heating or processing since they formed. Applying isotopic analysis to these CAIs, time zero has been calculated as 4,567 billion years ago.

A close up of a section of the Allende meteorite, which fell in Chihuahua, Mexico in 1969, reveals the CAIs that date the beginning of our Solar System to 4,567 billion years ago.

Composed of 66 antenna, the Atacama Large Millimeter Array (ALMA) in the Atacama Desert, northern Chile acts as a single telescope focussed on the light eminating from space. Scientists can use this light to study star formation and the early Universe.

The protoplanetary disk contained a wide diversity of material, including CAIs, metal grains, chondrules (millimetre-sized spheres of rock that formed from molten droplets), ice and complex organic molecules. At first, the material in the protoplanetary disk stuck together forming pebbles. Eventually, over thousands of years, the pebbles grew and attracted more material through gravitational forces and began to coalesce, in a process we call accretion, to form planetesimals up to hundreds of kilometres in size. These planetesimals (minute planets) eventually collided and formed the planets that we know of today. Telescopes based on the Earth, such as the Atacama Large Millimeter Array (ALMA) in the Atacama Desert, northern Chile, have observed these protoplanetary disks around other young stars, showing gaps in the disk where planets are forming and clearing the dust away in their orbit.

It is thought that Earth accreted within the first 70 million years of the Solar System forming. As there are very few rocks on Earth older than four billion years, it is challenging to fully understand the early Earth, but by using a combination of geochemical analyses and theoretical modelling scientists have begun to understand the evolution of our planet. Perhaps the most significant event in the Earth's history was the Moon-forming 'Giant Impact' around 70–120 million years after the Solar System formed. Modern theories propose that a Mars-sized planetesimal, dubbed 'Theia', impacted into the early Earth, massively affecting our young planet. Much of Theia's mass was absorbed by the Earth, but a huge amount of material was thrust into space, forming a disk of debris that later accreted to form the Moon.

The Moon itself has a considerable influence on the Earth. We have the Theia impact, and the Moon it created, to thank for the 23.4° tilt of the Earth's axis, which is the reason why we have seasons. This axis has a slight wobble, called the axial precession and occurring on a 26,000 year cycle, which is stabilized by the gravitational effect of the Moon. Without it, the Earth's axis would move at a much faster rate, resulting in extreme climatic shifts. The influence of the Moon can also be seen in the life cycles of many organisms. For example, organisms living in tidal habitats are adapted to diurnal changes in seawater levels, as tides are controlled by the Moon's phases, and the annual mass spawning of coral coincides with the lunar cycle.

When it formed, the Earth had a primitive early atmosphere consisting of gases that accreted as it grew and were held together by the Earth's gravitational pull. However, as volcanic

The Moon formed when a Mars-sized planetesimal impacted the early Earth, producing a debris disk of material that coalesced and accreted together.

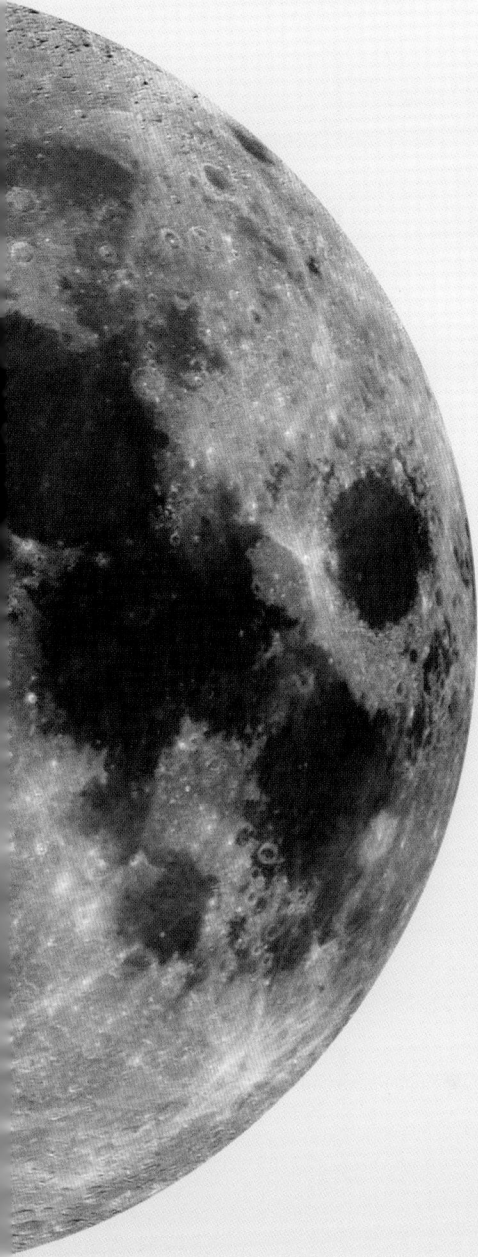

activity raged on the Earth's surface, a lot of carbon dioxide, hydrogen sulphide, methane and other volatiles were released, enriching and changing the composition of the atmosphere. After several million years, the surface cooled sufficiently for water to condense and collect. The origin of Earth's water is still debated by scientists. As temperatures in the inner region of the protoplanetary disk were too high for the formation of water–ice, it's likely that the materials from which Earth accreted were 'dry'. In contrast, beyond the 'snow line' – the distance from a star where temperatures are low enough for ices to form – ice crystals accreted together with the dust in the disk to form icy planets, asteroids and comets.

To explain the presence of water on the Earth, one common theory is that, after the surface had cooled, asteroid impacts deposited a late veneer of primitive meteoritic material onto the surface. Comets – large, icy objects made up of dust, rock and ice – were long thought to be a major source of Earth's water, but the ESA space missions Giotto in 1985 and Rosetta in 2004 proved otherwise when they measured cometary water and showed that it differed significantly from that on the Earth. However, more recent work on meteorites, such as Winchcombe (p. 22), as well as samples returned from the carbonaceous asteroids Bennu and Ryugu by the NASA OSIRIS-REx and JAXA Hayabusa2 missions, have shown that the isotopic composition of water bound up in clay minerals in these rocks has a striking similarity to the water in our oceans.

As our study of meteorites and samples returned by space missions develops, so does our understanding of the evolution of the Solar System, the origin of life on Earth and the potential for life elsewhere.

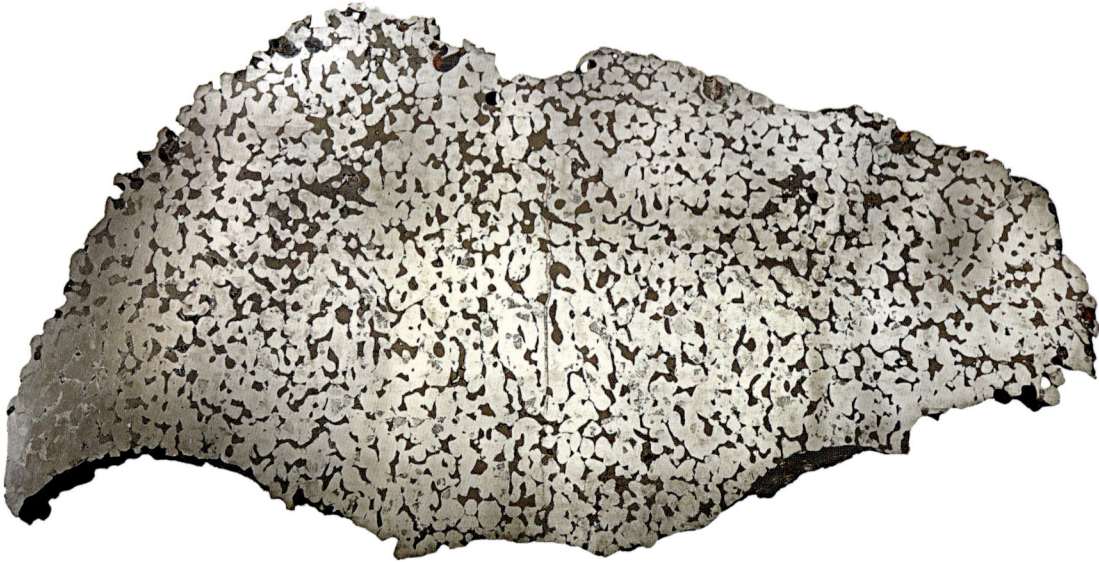

Mundrabilla meteorite

A slice of the iron meteorite found on the Nullarbor Plain, Western Australia in 1911. It is mostly made of around 65% iron–nickel metal and 35% of an iron-sulphide mineral called troilite. These types of meteorite are used to investigate planetary formation, in particular the formation and evolution of planetary cores in the early Solar System. Scientists also use the properties of iron meteorites to model what the composition and structures of moons and planetary cores may be like today.

Allende meteorite

This stone is from the largest carbonaceous chondrite ever found and fell on 8 February 1969 in Chihuahua, Mexico. It has an abundance of large calcium–aluminium-rich inclusions (CAIs), which are the first solids to form in the Solar System and allow us to measure the age of the Sun.

Parnallee meteorite and section

This meteorite fell on 28 February 1857 in Tamil Nadu, India. It is an ordinary chondrite, a type of meteorite that has not significantly changed through heating on its parent asteroid, and so it preserves a record of the materials from the early Solar System. The rounded objects visible on the surface, and clearly seen in the section above, are called chondrules. They formed from spherical molten droplets of silicate minerals around the young Sun before accreting together with iron-nickel metal grains to form chondritic asteroids – the source of this meteorite.

Estherville meteorite

This mesosiderite meteorite fell on 10 May 1879 in Iowa, USA. This type of meteorite is part of a complex and rare group of breccias – rocks made up of multiples of other rock types. It is thought that they formed by a collision(s) that resulted in the catastrophic fragmentation of the original parent bodies. The resulting asteroids (and then meteorites) formed from the remaiing material. Complex meteorites like Estherville reveal the active and chaotic environment of the early Solar System.

IRON METEORITES

Youndegin meteorite

This iron meteorite was found in Quairading Shire, Western Australia in 1884 and, in total, 3.8 tonnes of the fall have been recovered. Made predominantly of crystals of iron–nickel metal, it also contains inclusions of silicate rock similar to a group of stony meteorites or achondrites, suggesting they come from the same parent body. Meteorites like this are used to investigate planetary formation, particularly the separation, or 'differentiation' of heavier materials and lighter materials as larger bodies formed within the early Solar System. Iron sinks to form cores and the lighter, mineral-rich material rises to form the mantles and crusts of these minute planets known as 'planetesimals'.

Rowton meteorite

This iron meteorite fell in Shropshire, UK on 14 March 1876. It is the only known iron meteorite found in the UK and one of only 49 observed iron falls worldwide. The cut surface shows the crystals of iron–nickel metal, named the Widmanstätten pattern or Thomson structures. These would have formed as it cooled very slowly in the core of a large asteroid or small planet.

PALLASITE METEORITES

Esquel meteorite

This pallasite (stony-iron meteorite) was found in Chubut, Argentina in 1951. Pallasites were once thought to originate at the boundary between the core and the mantle of a large asteroid, where the metal and rocky components of the asteroid have separated in a process known as differentiation. However, a recent alternative hypothesis suggests the mixture of stone and metal may instead have formed by impacts between asteroids.

Krasnojarsk meteorite

One of the earliest meteorites to be scientifically described, Krasnojarsk was the first ever pallasite discovered, found in 1749 in Siberia, Russia. This particular specimen was presented by the Academy of Sciences in Moscow to the Natural History Museum, London in 1776 and was the first meteorite in its collection, before it was commonly accepted that meteorites have extraterrestrial origins.

EUCRITE METEORITES

Johnstown meteorite

This diogenite meteorite fell on 6 July 1924 in
Colorado, USA. Diogenites are plutonic igneous
rocks, meaning they form deep in the crust and
cool very slowly, as indicated by the large size
of the crystals in this specimen. Data collected
by NASA's Dawn spacecraft has enabled
scientists to link it to the asteroid 4 Vesta.

Stannern meteorite

This eucrite meteorite fell on 22 May 1808 in Moravia, Czechia. Eucrites – igneous rocks made predominantly from the minerals pyroxene and plagioclase feldspar – are very similar to terrestrial basalts, which form on Earth from cooling lava. LIke the diogenite meteorite opposite, it is linked to the asteroid 4 Vesta, which is one of the largest bodies in the main asteroid belt between the orbits of Mars and Jupiter. Studies of 4 Vesta and eucrite meteorites enable scientists to learn about the earliest stages of planet formation in the Solar System.

Winchcombe meteorite

As a small rocky body in outer space (meteoroid) entered Earth's space, it was observed, as a meteor, by multiple camera networks, just before 10pm on 28 February 2021. Scientists were then able to calculate the likely landing area and alert locals to be on the lookout for an unusual rock. The first fragments were found in the village of Winchcombe in the Cotwolds, England, which gives the meteorite its name. Winchcombe is the first carbonaceous chondrite, and one of only 19 meteorites to be seen to fall and be recovered, in the UK. Altogether, 602 grammes of the meteorite have been found in hundreds of fragments. The rapid collection and curation of Winchcombe make it one of the most pristine samples available for science, and initial studies of it support the theory that these types of meteorites were a major source of Earth's water.

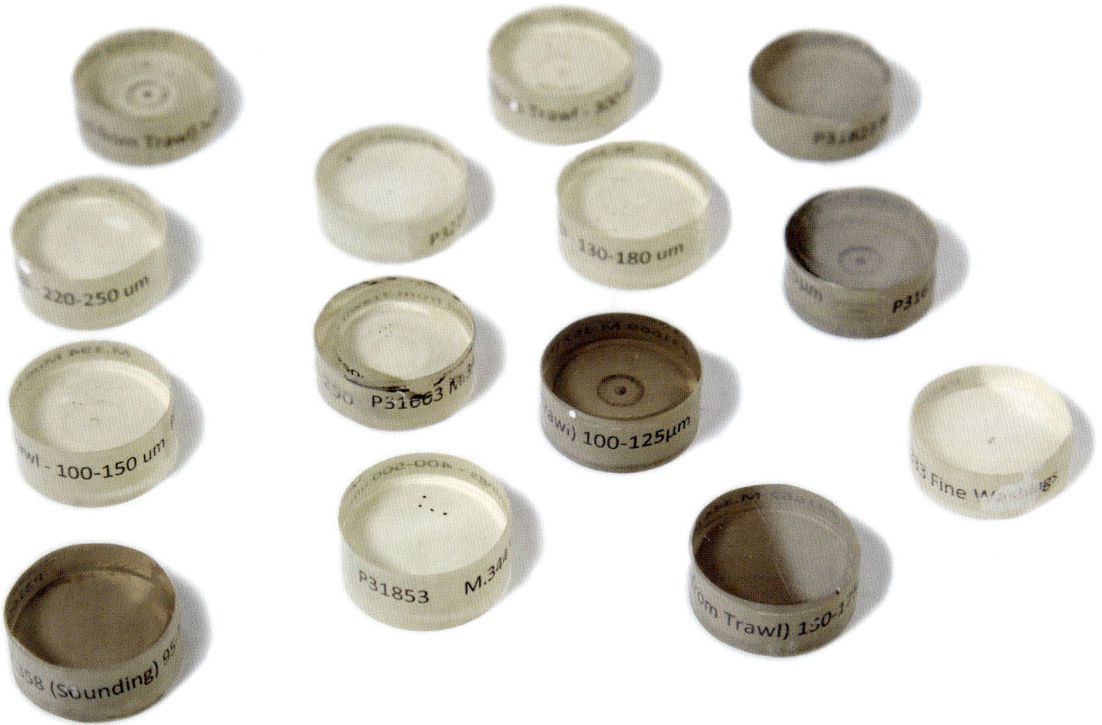

Micrometeorites

These polished blocks of epoxy contain micrometeorites that have been extracted from sediments collected by dredging the seafloor during the HMS *Challenger* expedition 1872–1876. The expedition circumnavigated the globe to investigate the deep sea, seafloor deposits and marine biology. Micrometeorites are important to science as they provide an estimate of the amount of extraterrestrial material that travels into the inner Solar System through time. They also allow us to investigate how this dust interacts with Earth's atmosphere and provide insight into the nature of different parent asteroids.

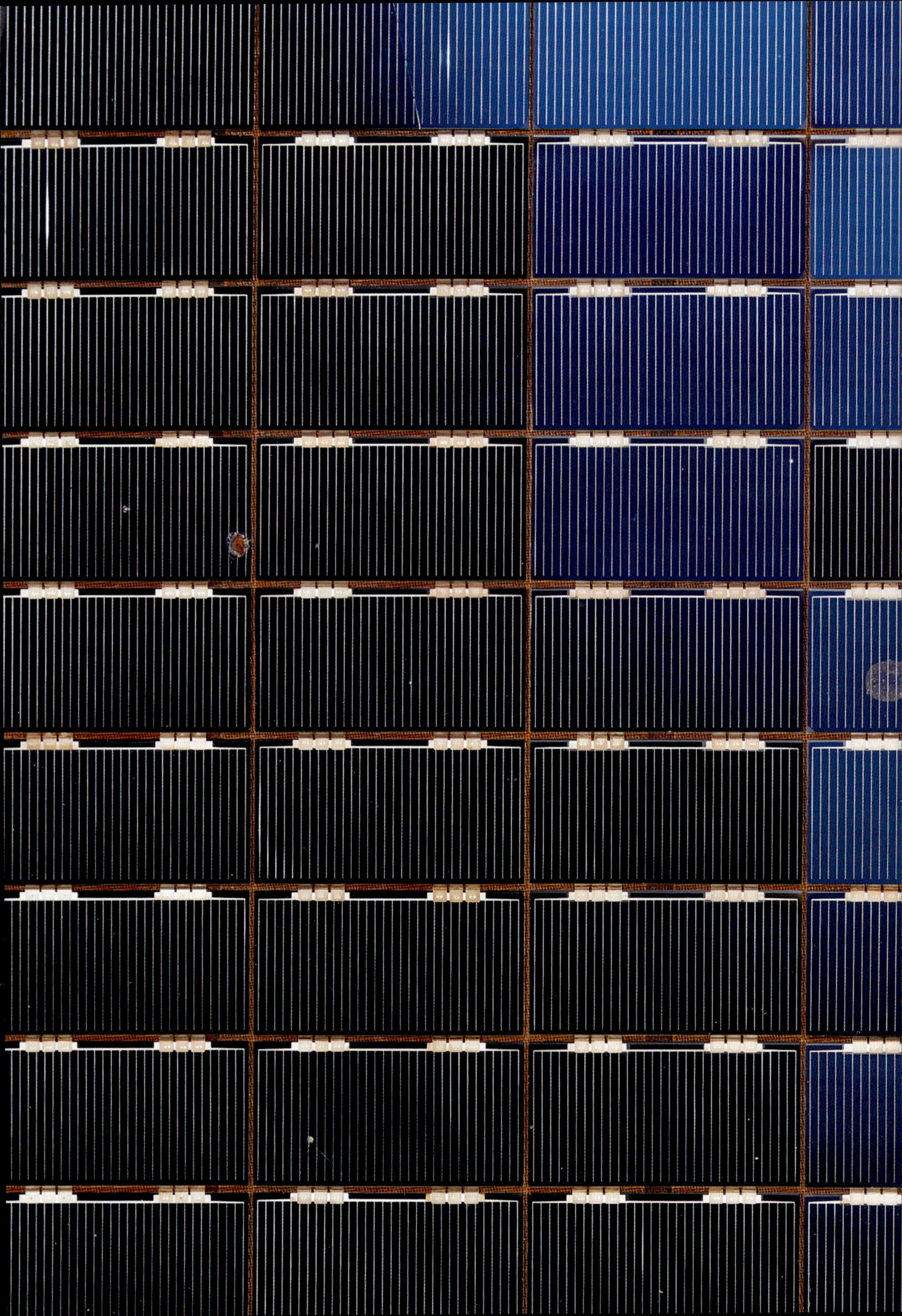

Hubble space telescope solar cells

This is a piece of NASA's Hubble Space Telescope Solar Array. It was retrieved from the telescope in low Earth orbit in 2002 after being damaged by numerous high-velocity impacts. Scientists have analyzed the composition of the material inside the tiny craters, to investigate the origin of these impacts and whether they are natural cosmic dust particles or whether they are artificial debris created by space missions. This research tells us about the type and abundance of extraterrestrial material arriving on Earth and helps plan future satellites and space missions.

Dresser formation

The Dresser Formation stromatolites are the oldest widely accepted direct evidence of life – almost 3.5 billion years old – and are found in an isolated part of Western Australia, several hours drive from the nearest town. These rocks are of similar age to the time when Mars was thought to be more suitable for life (i.e. more than three billion years ago) and are thought to have formed in comparable environments. Although they do not contain microfossils and have been gradually weathered, their two- and three-dimensional morphologies (shapes) include wavy layers and domes that are common in younger stromatolites; these features indicate the biological origins of the Dresser Formation stromatolites.

Stromatolite from the Transvaall Supergroup

Stromatolites are sedimentary layers formed by photosynthetic microorganisms, such as cyanobacteria, that cement sand and rocky materials together to form microbial mats. This one comes from the Transvaal Supergroup (a collection of two or more rock formations that share certain characteristics) in South Africa and is 2.5–2.6 billion years old. It formed on a shallow carbonate platform and dates from the period during which Earth's atmosphere began to be oxygenated. The oxygenation of the atmosphere was an unprecedented revolution in the Earth–Life system and made possible all subsequent steps in evolution. The rock record from this period of Earth history can be used to trace major environmental change over hundreds of millions of years, and stromatolites might provide clues as to how this oxygenation was initiated.

Gunflint chert

At the time of its discovery in the 1950s, the North American Gunflint Chert provided the oldest known evidence of life, in the form for filamentous and spheroidal microfossils – it is 1.88 billion years old. These well-preserved microscopic fossils have been interpreted as examples of cyanobacteria, the organisms likely responsible for the oxygenation of Earth's atmosphere. Although it at first appears a rather unremarkable and featureless specimen, the Gunflint Chert is among the most historically significant rocks in the context of advancing our knowledge of early life on Earth.

Stromatolite from the Biwabik Iron Formation

In some cases, such as with this 1.88 billion-year-old stromatolite from the Biwabik Iron Formation in Minnesota, USA, microbes can construct complex layered column-like structures that become preserved in rocks (the vertical features visible in this specimen). These structures are common in shallow-water microbial mats and provide crucial macroscopic indications – visible to the human eye – of biological activity. The rich red, orange and brown colours of this specimen are due to layer replacement of stromatolite layers with iron-rich minerals.

Asteroids

Ashley King

Asteroids are the rocky and metallic debris left over from the formation of the Solar System around 4.6 billion years ago. Ranging from only a few metres up to about 500 kilometres (310 miles) in diameter and largely unmodified by geological processes, asteroids can tell us when, where and how planetary bodies like Earth and Mars formed. Today, most asteroids reside in the approximately 225-million-kilometre (140-million-mile) wide main asteroid belt between the orbits of Mars and Jupiter. However, there are also populations of asteroids whose orbits pass relatively close to Earth (near-Earth asteroids) or follow a similar path to Jupiter around the Sun (Jupiter Trojans), pointing to an intricate dynamical history shaped by collisions and gravitational interactions with the Solar System's giant planets.

The physical and chemical properties, such as size, mass, orbit and albedo (how much light they reflect), are known for roughly one million asteroids, and there are many more that are yet to be characterized in this way. Over 70% of asteroids have very dark surfaces, probably because carbon makes up several percent in weight of their composition. Classified as carbonaceous (C-type) asteroids, many of these small bodies are also thought to contain hydroxyl and water molecules, not in liquid form but rather bound within the structure of hydrated minerals.

Organic compounds, one of the essential ingredients needed for life, were first directly detected on carbonaceous asteroids in the late 1980s. Using powerful telescopes and spacecraft that analyze spectral characteristics of the sunlight reflected back from an asteroid's surface, astronomers have identified features on asteroid surfaces that suggest the presence of organic compounds. For example, on the surface of the carbonaceous dwarf planet Ceres, which was visited by NASA's Dawn mission, organic compounds were even found concentrated in the youngest impact craters. However, given how little light bounces back from dark carbonaceous asteroids, the study of organic compounds on their surface using ground- and space-based observations remains challenging.

Fortunately, fragments of carbonaceous asteroids naturally arrive on Earth as carbonaceous chondrite meteorites. We can study these rocks in the laboratory at a level of detail not yet possible with telescopes and space missions. Carbonaceous chondrites are rich in organic matter including hydrocarbons, and alcohols and a large number of amino acids, which are key ingredients in the development of proteins. The complexity of the organic matter in carbonaceous chondrites suggests that it evolved through a wide range of processes and different environments including, in some cases, contamination when it entered Earth's atmosphere.

To overcome this issue of contamination, JAXA's Hayabusa2 and NASA's OSIRIS-REx missions recently collected and returned to Earth samples of the carbonaceous near-Earth asteroids Ryugu and Bennu, respectively. Unlike the meteorites in our collections, the Ryugu and Bennu samples have never been exposed to the terrestrial environment. This means that for the first time ever, we are getting a pristine view of the full inventory of organic materials that were available at the start of the Solar System.

Conditions on carbonaceous asteroids were never favourable for living organisms – they have no atmosphere, little gravity and lack a stable source of heat. But with an abundance of carbon, nitrogen and water, they acted as important geochemical factories for the synthesis of complex organic compounds. As planetary bodies accreted in the early Solar System, carbonaceous asteroids played a crucial role in delivering prebiotic organic molecules to their surfaces where, together with the other essential ingredients for life of water, energy and 'just right' conditions, they took the first steps towards the formation of life.

Asteroids are the leftover building blocks of Solar System formation. Most can be found in the main asteroid belt between the orbits of Mars and Jupiter. However, some asteroids were scattered towards the inner Solar System by impacts and gravitational interactions with the giant planets and have orbits that pass relatively close to Earth. These near-Earth asteroids can be divided into distinct populations, including the Amors, Atens and Apollos.

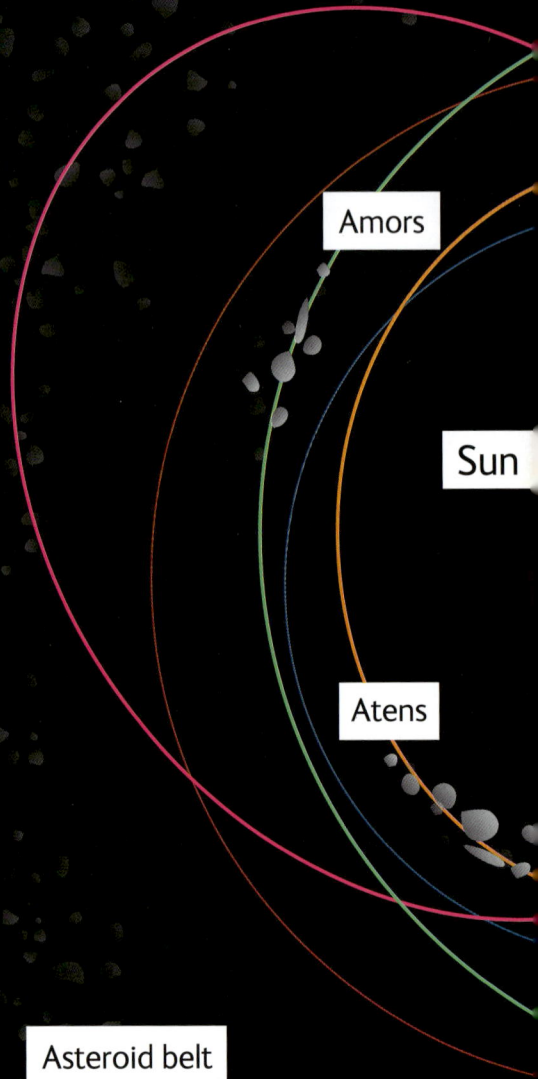

Amors

Sun

Atens

Asteroid belt

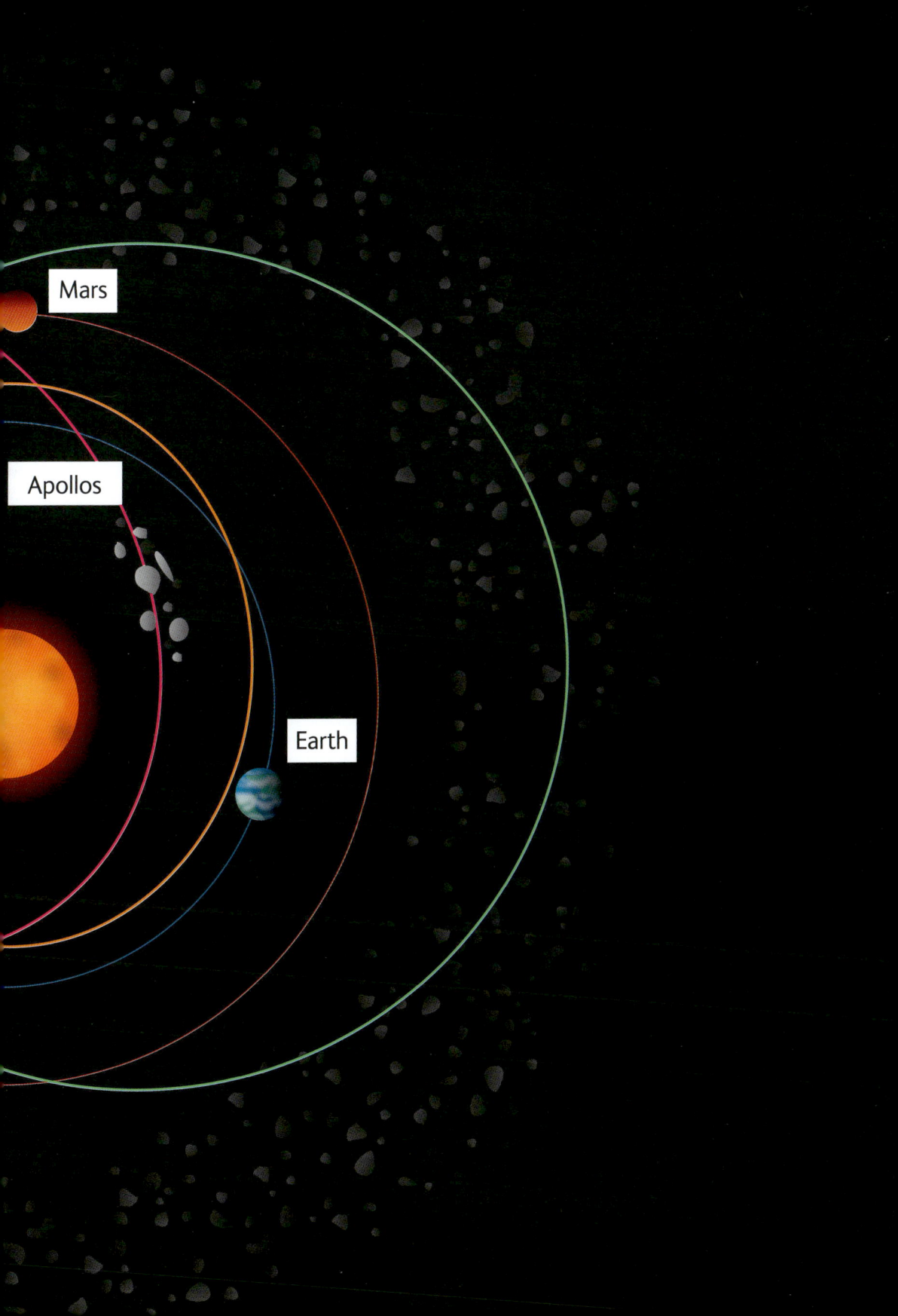

Mars

Apollos

Earth

MINERVA-II1 JAXA

In 2018, MINERVA-II1 (2nd generation MIcro- Nano Experimental Robot Vehicle for Asteroid) became the first rovers to be deployed onto the surface of an asteroid as part of JAXA's Hayabusa2 mission to the near-Earth asteroid Ryugu. The rovers, each weighing about 1 kilogramme (2 pounds), carried a series of cameras, temperature sensors and an accelerometer (for measuring the acceleration due to the pull of Ryugu's gravity). Gravity on the surface of Ryugu is so weak that a rover with wheels would float upwards when it moved. So the MINERVA-II1 rovers used a hopping mechanism to explore multiple areas, sometimes spending nearly 15 minutes above the asteroid surface before landing.

Pieces of the Ryugu asteroid

These are rocky fragments and particles from
the asteroid 162173 Ryugu, collected by the
JAXA Hayabusa2 mission, and returned to the
Earth on 5 December 2020. It was the first
sample returned from a carbonaceous asteroid,
targeted for its primitive composition and
therefore insight into the types of materials
and processes of the early Solar System,
including water and organic compounds
relevant to the origin of life. Hayabusa2
deployed a lander, as well as the two rovers,
to explore the asteroid surface. The mission
collected 5.4 grammes (0.2 ounces) in two
samples from Ryugu – one from
the surface and another from
the subsurface, using an
impactor to create
a crater on the
asteroid.

Ivuna meteorite

This meteorite fell in Mbeya, Tanzania on 16
December 1938. Ivuna is the type specimen of
the CI carbonaceous chondrites – the most
chemically primitive meteorites known, with
compositions similar to the solar photosphere,
the visible surface of the Sun. They contain
abundant phyllosilicates, or clay minerals,
which formed through the original minerals
being altered by water on the parent asteroid.
These meteorites are exceptionally rare and
only ten are known to science.

NASA Genesis Collector Array

Launched in 2001, NASA's Genesis mission
was designed to collect samples of the solar
wind and then return them to Earth. The
spacecraft had five collector arrays consisting
of hexagonal wafers made of high-purity
materials including aluminium, sapphire, silicon
and gold. Each array was exposed to the solar
wind, with billions of atoms of solar particles
implanted below the surface of the wafers.
Despite a hard crash landing in the Utah
desert, USA in 2004, laboratory analysis of the
collector arrays has refined our understanding
of the Sun's elemental abundances, giving us
new insights into the starting composition of
the Solar System.

OSIRIS-REx

Robotic arm

The Touch-and Go Sample Acquisition Mechanism (TAGSAM) is a robotic arm about 3.4 metres long on NASA's OSIRIS-REx spacecraft, which returned samples of the near-Earth asteroid Bennu to Earth in September 2023. During contact with the asteroid surface (the 'TAG manoeuvre'), which lasted around six seconds, the round sampler head used nitrogen gas to lift dust and small pebbles from the surface to inside the head. The sampler head also had 24 circular contact pads comprised of fine loops that captured small particles from the top layers of Bennu's surface.

OSIRIS-REx

Capsule

Shortly after TAGSAM's successful contact with the surface of the near-Earth asteroid Bennu in October 2020, the sampler head was stowed inside the sample return capsule (SRC). The SRC – 80 centimetres (31 inches) in diameter and 50 centimetres (19 inches) in height – preserved the asteroid sample until the 24 September 2023 when, at a distance of about 100,000 kilometres (62,138 miles), the OSIRIS-REx spacecraft ejected the capsule into the Earth's upper atmosphere. The outer layers of the SRC were specially designed to resist temperatures of nearly 3,000°C (5,432°F) during atmospheric re-entry.

Capsule parachute

After release from the OSIRIS-REx spacecraft, the SRC coasted for about 4 hours before travelling through the Earth's atmosphere, where it reached speeds of just over 43,450 kph (27,000 mph). The main parachute, which was deployed at an altitude of around 3 kilometres (2 miles) slowed the SRC down to about 17 kmph (11 mph), bringing it to a soft landing in the Utah desert, USA, 13 minutes after re-entering Earth's atmosphere. The SRC was recovered from the desert by helicopter and transported to a clean laboratory little over an hour after landing back on Earth.

OSIRIS-REx

Bennu sample

This is the sample of the asteroid 101955 Bennu in the TAGSAM sampler head, collected by the NASA OSIRIS-REx mission and returned to the Earth on 24 September 2023. Telescopic observations of the asteroid indicated a primitive, carbonaceous composition with water-bearing minerals, suggesting Bennu could contain compounds relevant to the origin of life and habitability of planetary bodies. The mission aimed to return and analyze the carbonaceous material to study the nature, history and distribution of the minerals and organic compounds present, as well as to characterize the asteroid and its geological history. The mission returned 121.6 grammes (4.3 ounces) of material from one site called Nightingale, and preliminary analyses confirm the presence of an abundance of hydrated minerals and organic compounds.

Mars

Peter Grindrod

Mars and Earth were probably very similar after they formed in the young Solar System around 4.6 billion years ago. Both young planets probably had carbon dioxide-rich atmospheres but followed divergent evolutionary paths. The Earth ultimately became a world of plate tectonics, oxygen and life whereas, at about the same time, Mars lost much of its atmosphere, water and, possibly, its habitability.

Early in its history, about 4 billion years ago, Mars almost certainly had large volumes of liquid water on its surface, and was probably much warmer than it is today. Together, these provided a habitat that could have been favourable for life. However, Mars began to lose its atmosphere at approximately the same time as the release of gas from volcanic eruptions on its surface created an acidic water environment, probably around 3.5 billion years ago. The speed of this acidic transition could have determined whether life could have adapted to the changing conditions. As the atmosphere thinned, some of this water was lost to space, but large volumes were stored as subsurface ice and in hydrated minerals.

For about the last three billion years, Mars has been a relatively dry and cold planet, but it is thought that periodic

A global mosaic of Mars, made using Viking Orbiter images. The 3,000 kilometre (1,864 mile) long canyon system, Valles Marineris, is visible in the centre of the image.

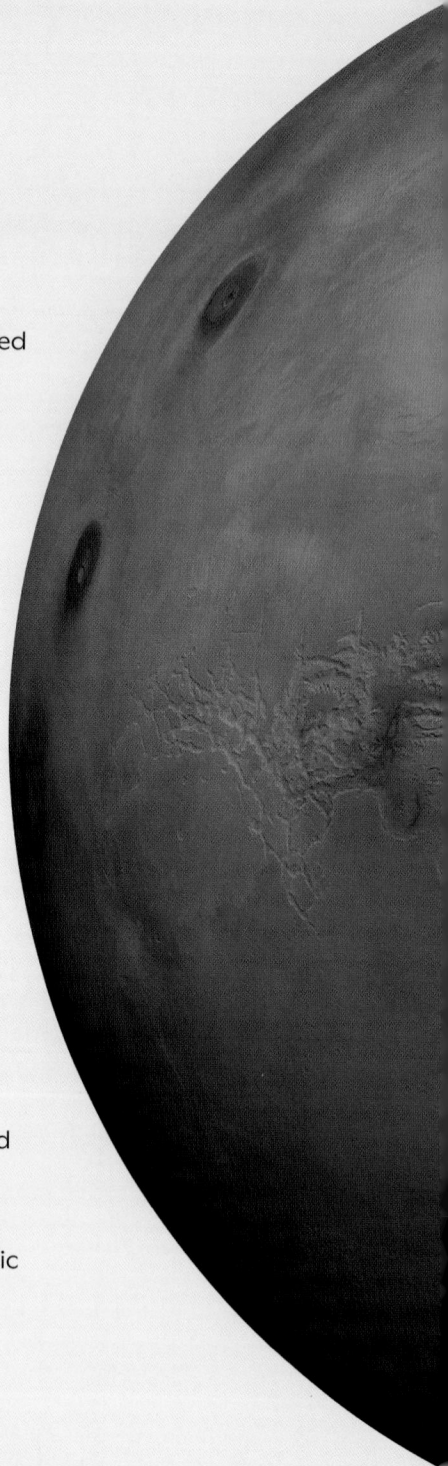

changes in the tilt axis of Mars have caused global climate change and the redistribution of water over its surface during at least the last few million years, and probably much longer. It is also possible that there are brief periods of liquid water on its surface today over much shorter, seasonal timescales.

Our knowledge of the evolution of Mars has come from a concerted effort to explore it, with a series of evermore advanced and capable spacecraft. Much of this understanding has arisen from data returned from spacecraft orbiting Mars. Gradually, the resolution and capability of instruments has improved, and there have even been a number of successful landings on the surface, so that recently we have begun to understand the planet from close up. Through missions such as NASA's Curiosity and Perseverance rovers, we have robotic geologists exploring the surface of Mars, returning data on a daily basis.

At present, the only samples we have of Mars are Martian meteorites. These samples are produced when a large impactor hits Mars with enough energy to launch some material into space so that it escapes Mars' gravitational influence. Some of these samples have then crossed the orbit of Earth where they fall to the surface as meteorites. By comparing detailed studies of the composition of the meteorites on Earth, to results from Mars missions, we can identify these samples as originating from Mars. To date, approximately 200 individual meteorites have been found on Earth that are thought to come from about ten impact events on Mars. Identifying the exact source region on Mars is difficult but is crucial to us not only gaining context for the meteorite samples, but also to allowing us to better understand the evolution of Mars.

Although efforts to improve our knowledge of the potential source craters for the Martian meteorites continue, missions that return samples from Mars, selected based on the habitability of an area, have long been a priority. At present, the Perseverance rover is collecting cores from rocks in the Jezero Crater on Mars, which will be returned to Earth on a future mission. These cores will undoubtedly become some of

This Perseverance rover (about 3 metres tall and 2 metres wide) selfie image is a mosaic of 62 individual images taken by the WATSON instrument on the arm of the Perseverance rover. It was taken on the 1,218th day (or 'sol') of the mission, at an area with a nickname of Cheyava Falls.

the most studied samples in planetary science and provide a new paradigm in our understanding of Mars. In addition, the European Space Agency ExoMars Rosalind Franklin rover will land in the Oxia Planum region of Mars in about 2030, with the specific goal of drilling 2 metres (6½ feet) into the surface and looking for evidence of past life.

This combined effort of studying Mars in situ with rovers and spacecraft, and with returned samples in laboratories on Earth, could well provide the first evidence of life beyond Earth.

Phoenix lander robotic arm

The Phoenix lander was a mission to a polar
region of Mars, which operated on the surface
for about 8 months in 2008. One of the main
reasons that Phoenix landed at such a high
latitude, about 68 degrees north, was to search
for water ice in the subsurface. The lander
used this 2 metre (6½ feet) arm to scoop up
the material at the surface, for analysis in the
lander. The small trenches dug by the arm
revealed water ice just a few centimetres below
the surface. Unlike Phoenix, which sampled
in situ (and the Curiosity rover of 2011 which
also drilled and analyzed samples in situ),
Perseverance is collecting samples that it is
planned will be returned to Earth.

PERSEVERANCE

Rock sample tubes

The Perseverance rover landed in the Jezero impact crater on Mars in February 2021. Since then, it has been exploring the surface, driving nearly 30 kilometres (18½ miles) as of 2024, and collecting rock samples along the way. It is planned that these samples will be returned to Earth by a future mission, and have been selected for their importance in the geological history of Mars and also the search for life. The 43 sample tubes are made mostly of titanium and are designed to preserve the rock samples until they have been returned safely to Earth. It is vital that there is no contamination, and so the seals that are inserted after rock collection are a crucial part of the mission design. Before any space mission takes place, many equipment tests take place on Earth. For the Perseverance rover, a key part of the mission is using a drill to collect a core of a rock. There have been many tests of this drill and core system on Earth to not only ensure that the overall method works, but to also determine the best way of sampling different rock types.

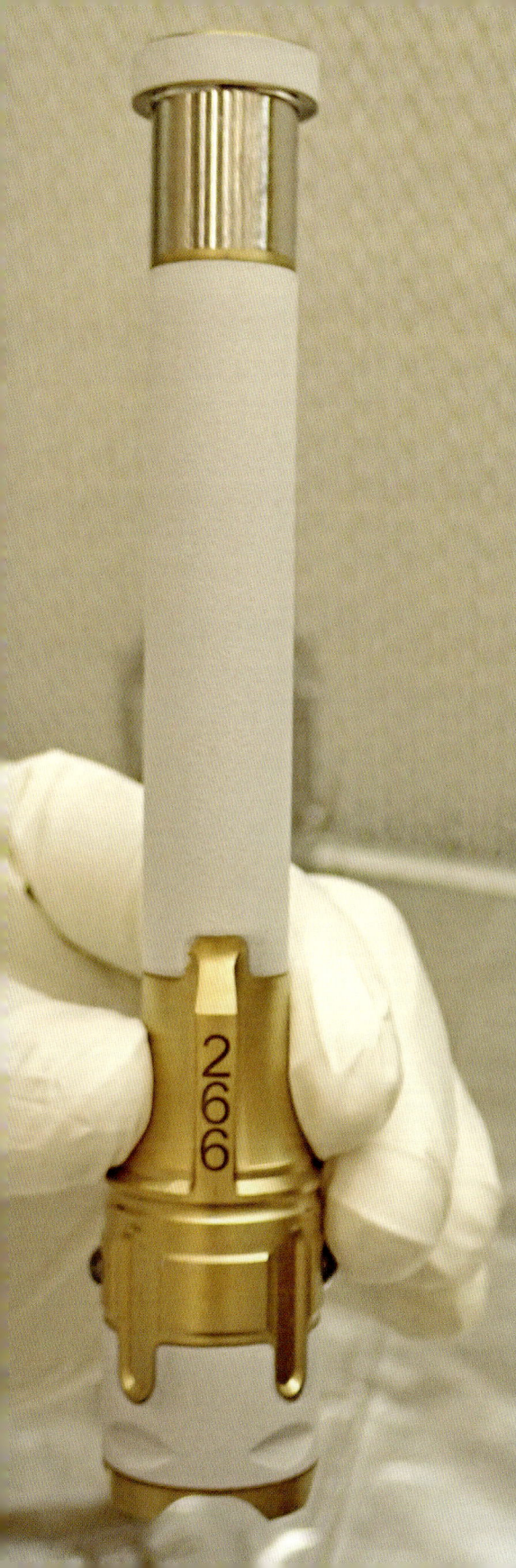

PERSEVERANCE

Drill

The main goal of the Perseverance rover is to collect rock core samples using a drill mechanism. This drill has to be designed and tested to withstand not only the violent conditions experienced during launch on a rocket from Earth, but also the landing on Mars. Moreover, the drill has to work perfectly for several Earth years under the freezing conditions on the surface of Mars.

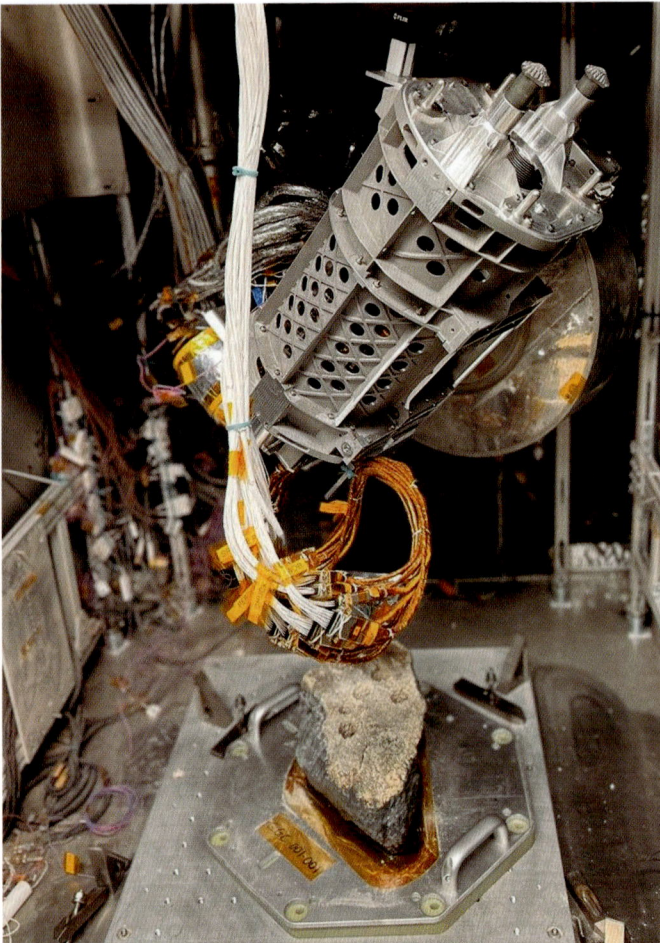

Drill stabliliser

Before the Perseverance rover can begin to drill on Mars, it must ensure that it has a good contact with the rock that is being sampled. This stabiliser mechanism allows the drill to preload onto the surface, and prevents the drill bit from moving around when drilling is taking place.

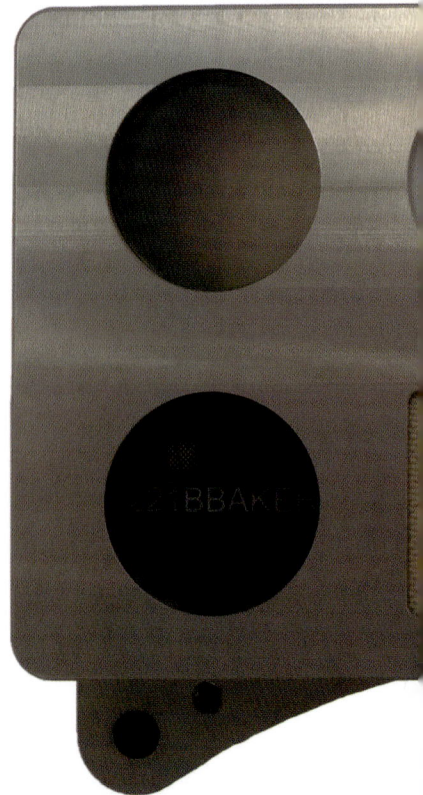

PERSEVERANCE

Core drilling bits

The tip of the drill mechanism on the Perseverance rover is designed both to drill through different rock types and to collect a core sample about 13×60 millimetres (½x2¼ inches) in size. The rock core is then transferred to a sample tube and sealed, ready for the journey to Earth.

SHERLOC calibration target

The SHERLOC calibration target is part of the scientific payload – instruments that have the specific purpose of collecting data and performing experiments – on the Perseverance rover. Situated on the front of the rover, the target is made up of 11 different 'target' materials, which the scientific team use to test the operation of the SHERLOC instrument. Some of the targets are prototype spacesuit materials and are being tested to see how they withstand the harsh, Martian surface environment.

PERSEVERANCE

Martian meteorites

Samples from the Sayh al Uhaymir 008 (SaU 008) Martian meteorite, found in Oman in 1999. This type of meteorite is known as a shergottite and is composed of basaltic rock, which is extremely common on Mars. A portion of this meteorite has 'returned home' to Mars on the Perseverance rover as part of the SHERLOC calibration target (*see* p.53).

EXOMARS ROSALIND FRANKLIN

Rover

This mission has the main goal of searching for life on Mars. Due to launch in 2028, the rover –about 2 metres (6½ feet) tall with solar panels 2.5 metres (8 feet) wide – will land in the Oxia Planum region of Mars, which is rich in phyllosilicate minerals that suggest there was a water-rich environment about 3.5 to 4 billion years ago. The rover will collect and analyze the first deep samples on Mars, returning drill cores from 2 metres (6½ feet) below the surface, where any possible evidence for life is likely to be better preserved. The samples will then be analyzed by sensitive instruments within the body of the rover.

EXOMARS ROSALIND FRANKLIN

PanCam colour calibration instrument

It is vital to ensure that all instruments operate well throughout their lifetime when they are on Mars. It is particularly important with image-taking instruments that colours are calibrated when they are on Mars because of the different lighting and atmospheric conditions that are experienced. These calibration targets – the black area and its components shown below – will be imaged regularly to ensure that the PanCam instrument is properly calibrated. The imaging will also study the accumulation of dust over time, which is important to know for solar-powered missions. An accumulation of dust on the solar panels means there is less power available to the rover, which means it can't do as much, and might even end the mission eventually.

PanCam filter wheel mechanism

The Panoramic Camera, or PanCam, instrument is the eyes of the ESA ExoMars Rosalind Franklin rover. The camera takes images in different wavelengths, or colours, which can be used to understand the composition of the rocks. The camera is instructed to take an image, and then that image is returned to Earth. The team back on Earth will then spend some time (probably several days) using that image and other data to decide whether to drill there or not. This process will be vital in selecting potential targets to be sampled by the deep drill.

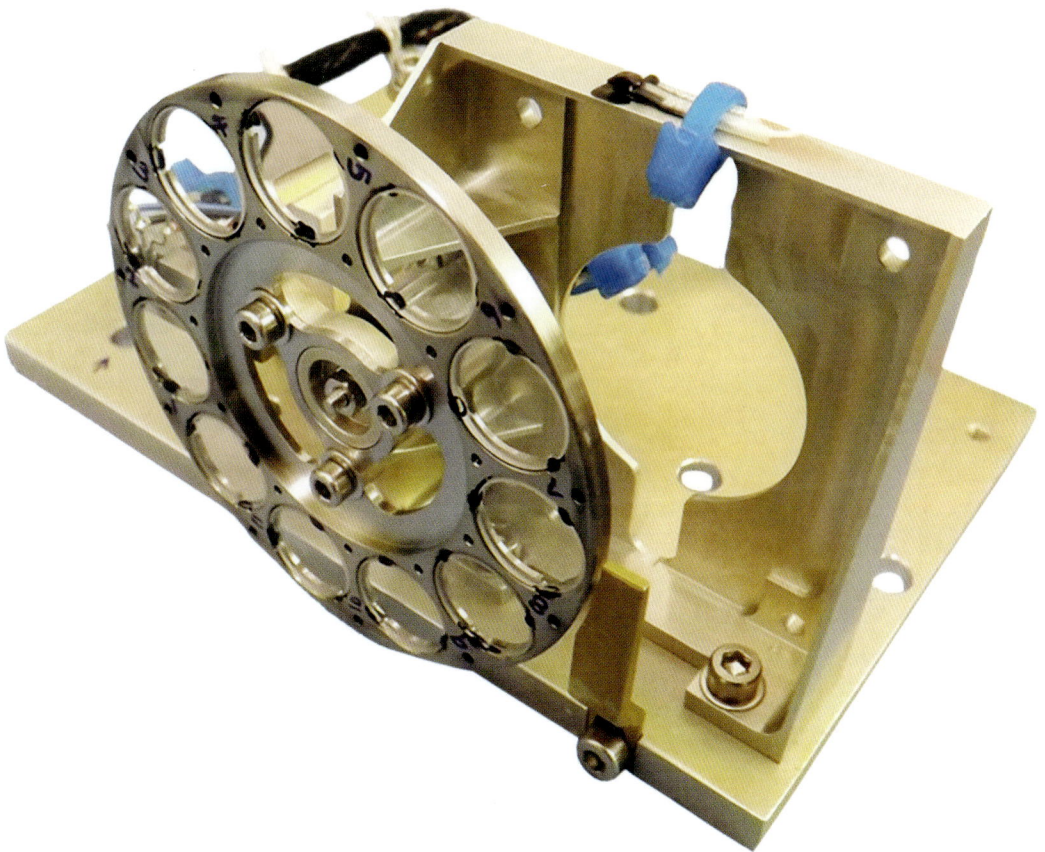

MARTIAN METEORITES

Amgala meteorite

This Martian shergottite was found in 2022 in Western Sahara. Altogether, almost 35 kilogrammes (77 pounds) was found in many pieces. Martian meteorites account for only 0.5% of all known meteorites. Shergottites are of igneous origin – they crystallize from magma – and provide key insights into our understanding of the geological evolution of Mars and show that Mars had active volcanism as recently as a few hundred million years ago.

Northwest Africa 11220 meteorite

This meteorite was classified at the Natural History Museum, London in 2017 and is one of multiple stones known as the Northwest Africa 7034 pairing group (when lots of the same meteorite type are found in close proximity this is known as a pairing group), found in Rio de Oro in Western Sahara. This group is exceptional amongst Martian meteorites. It defined a new category of meteorites, because it contains many different types of rock and has a high water content. It contains some fragments as old as 4.48 billion years.

MARTIAN METEORITES

Tissint meteorite

This shergottite was observed falling on 18 July 2011 and is the most recent Martian fall. It landed in the Guelmin-Es Semara region of Morocco. It formed from a magma under the Martian surface around 540 million years ago. The thin black glassy veins on the broken surface formed by shock, very probably when the sample was ejected from the Martian surface. The black glass contains small vesicles, or bubbles, of trapped gas. The composition of this gas can be analyzed, and it has the same carbon and nitrogen isotopic composition as the Martian atmosphere.

Allan Hills meteorite

Allan Hills 84001 is a Martian meteorite found during a snowmobile ride on 27 December 1984, in the Far Western Icefield of Allan Hills, Antarctica. It was immediately recognized as the most unusual rock collected during that field season, and is the only known orthopyroxenite (a type of igneous rock) from Mars, with the oldest crystallization age (between 4 and 4.5 billion years ago) of any Martian meteorite known thus far. The rock shows evidence of carbonate formation, indicating alteration by water some time after it initially formed. ALH84001 was also the subject of some controversy when, in 1996, there were claims that researchers had found fossils of bacteria-like lifeforms. These findings have since been discredited but did attract a lot of attention to the burgeoning science of astrobiology.

Nakhla meteorite

This meteorite fell as part of a shower of stones on 28 June 1911, near Alexandria in Egypt. Nakhla is one of only five observed falls from Mars. These falls are particularly important as they preserve the original Martian signatures, as opposed to meteorites that are found on the Earth's surface, which may have experienced significant weathering whilst on Earth. In the 1970s, scientists from the Natural History Museum, London described veins of clay minerals in Nakhla, indicating the presence of past water on Mars. This indicates that in the past, Mars was warm and wet, possessing a more habitable environment than today.

PATAGONIAN STROMATOLITES

Stromatolite on substrate

This stromatolite comes from the drying Carri Laufquen lake system in Patagonia, and is perhaps only 8–14,000 years old. It is growing on a pebble of basalt that has provided both a substrate to stabilize the growth of the stromatolites and key nutrients, including metals, that are essential to its metabolisms. It grew in a basalt-hosted lake, which is believed to be very similar to ancient habitable environments on Mars. Did similar microbial ecosystems also exist on Mars billions of years ago?

Layered stromatolite

In this Patagonian stromatolite, you can see sequential layers of microbial growth that may reflect growth patterns and seasonal or regional climatic fluctuations. Abundant organic microfossils, primarily cyanobacteria, are preserved within it. By studying the relationships between organic materials and carbonate minerals in such specimens, we can understand how microfossils become preserved over billions of years in Earth's oldest rocks, and develop potential hypotheses for how similar preservation might have occurred on Mars.

Domed stromatolite

This top-down view of a Patagonian stromatolite shows numerous small dome-like structures built by cyanobacterial communities. Stromatolites such as this are thought to be good targets for developing life-detection strategies for Mars exploration. They are dominated by microbial communities (consistent with ideas that the only types of life that might have emerged on Mars would have been microbial) and have carbonate compositions similar to those seen at the margins of Jezero crater on Mars. We can therefore use these types of specimens to develop life-detection strategies on Mars, helping to guide Mars rover exploration and define the mineral compositions that might help preserve traces of life.

Icy worlds

Caroline Smith

In 1609 the famous Italian polymath, Galileo Galilei (b. 1564) was the first scientist to use the newly invented telescope to view the night sky. He observed our Moon's cratered and mountainous surface and also turned his telescope to the largest planet in our Solar System, Jupiter, and was amazed to see that it appeared to have four 'stars' orbiting it. He recorded his observations of the Moon and the four objects around Jupiter in his book *Sidereus Nuncius* ('Starry Message') in 1610. A German astronomer, Simon Marius, was also using his telescope to study Jupiter at around the same time as Galileo, and it is the names he gave the moons that are familiar to us today – Ganymede, Callisto, Io and Europa. These moons are also known as the Galilean moons in recognition of Galileo's discovery.

From those first observations by pioneering astronomers in the early 17th century, hundreds of moons have been discovered orbiting other planets in our Solar System, particularly those planets in the outer Solar System. Many of these discoveries have been made through technological advances in telescopes, including the Hubble Space Telescope, and through data acquired by the first space missions to explore the outer planets – NASA's Voyager 1 and 2 probes between 1979 and 1989. As of late 2024, the current tally stands at Jupiter (95), Saturn (146), Uranus (27), Neptune (14) and even tiny Pluto has five moons.

As well as revealing the presence of many more moons, technological advances in telescopes and space exploration also made some unexpected and startling discoveries. Instead of being barren, inhospitable objects orbiting their parent planets (as we think of our own Moon), some of these moons appear to be dynamic bodies, showing surface features such as volcanic landscapes and ice sheets like those on Earth, and 'geysers' of water erupting into space.

Using data collected from space exploration missions, scientists now think that six moons, orbiting planets in our Solar System, may have oceans beneath their outer crusts. Tantalisingly, Jupiter's moons Ganymede, Callisto and Europa, Saturn's moons Titan and Enceladus, and Neptune's moon Triton may all have the right conditions for life.

Six moons have proved particularly interesting in the search for habitable environments and the search for life – Jupiter's moons Ganymede, Callisto and Europa, Saturn's moons Titan and Enceladus, and Neptune's moon Triton. These moons are thought to have many of the right ingredients and conditions for life, for example, liquid water below a protective crust, a heat source and the presence of organic molecules. This knowledge has focussed efforts to study these fascinating bodies in the greatest detail yet, using new and exciting exploration missions and the advanced James Webb Space Telescope.

Ganymede is Jupiter's largest moon, the largest moon in the Solar System and the only moon to generate its own magnetic field, which is thought to be caused by it having a liquid iron-rich core, similar to the Earth. Ganymede is also thought to have a thin atmosphere that contains oxygen, and its surface crust shows two distinct terrains. The dark regions are thought to be made up of icy rock, and the brighter regions predominantly water-ice. Beneath the crust, Ganymede is thought to have a large ocean. At an estimated 97 kilometres (60 miles) thick, this is about 10 times deeper than our own Earth's oceans.

Callisto is Jupiter's second largest moon and is larger than our
own Moon. Its surface is completely covered in impact craters and is
the most heavily cratered object in our Solar System. This indicates
that its surface is both very old (perhaps the oldest in the Solar
System) and that Callisto has never had any geological activity like
volcanism. Like Ganymede, the surface shows brighter areas, which
are thought to be made up of water-ice, and darker areas thought to
be made of an ice and rock mixture. It is not known whether Callisto
has a subsurface ocean and this is still a topic of scientific debate.

Europa is the smallest of the Galilean moons. Its surface is entirely
covered in a thick crust of ice, and the presence of cracks and long,
linear ridges in this icy crust, indicate that there is an active and
dynamic environment below. Evidence suggests that there is a vast

liquid ocean beneath the icy crust, potentially in contact with a rocky seabed. The interaction between water and rock could create the chemical ingredients necessary for life. Observations of Europa by the Hubble Space Telescope and the Galileo spacecraft have indicated that large eruptions of water vapour from the surface sometimes occur – these are known as cryogeysers.

ESA launched its Juice mission in April 2023, to investigate Ganymede, Callisto and Europa, and NASA launched its Europa Clipper mission in October 2024 to investigate Europa. These missions are due to arrive at their destinations in 2030 (Europa Clipper) and 2031 (Juice) to start their work, using their diverse suites of scientific instruments to study the physical and chemical properties of the moons and their surrounding environments.

Saturn's moon Titan is also a key object of interest in understanding the conditions needed for life to start and, perhaps, even find evidence of life on Saturn's largest moon. Compared to Ganymede and Europa, Titan has a very different environment. It has a thick atmosphere composed of nitrogen, methane and hydrogen, all key components to produce complex organic molecules that are crucial for life. Indeed, the conditions on Titan today are thought to resemble those on the early Earth. Titan's surface is covered by lakes and seas of liquid methane and ethane. These hydrocarbons, although very different from water, could potentially serve as a solvent for life. Life forms on Earth rely on water as a solvent because of its ability to dissolve a wide range of compounds, but methane and ethane could serve a similar role on Titan.

NASA has plans for an innovative and ambitious mission to visit Saturn's moon Titan and to land a car-sized vehicle with eight helicopter-like rotors on the surface. The Dragonfly rotorcraft lander is due to be launched in the summer of 2028 and arrive on Titan in 2034. With its ability to fly to and land at different sites on Titan's surface, the Dragonfly mission aims to investigate and characterise the composition of the atmosphere and surface, assess the habitability of the sites, investigate the chemical processes that could lead to life, and even look for signatures of life existing on Titan today.

This image of Jupiter's moon Europa was captured by the JunoCam on board NASA's Juno spacecraft during the mission's close flyby in September 2022. In the image, north is to the left and, like the Earth's Moon, one side of Europa always faces towards Jupiter. This is the side of Europa that is seen in this image. Europa's icy surface is crisscrossed by cracks, ridges and bands, indicating a dynamic environment below.

LUNAR METEORITES

Ajdabiya meteorite

This lunar meteorite was found in 2021 in Libya. Like most lunar meteorites from our Moon, it is a breccia – a rock made up of many rock fragments. Breccias are common on the Moon, owing to the frequent impacts experienced on the surface. Many lithologies, or rock types, including basalts from volcanic activity, and anorthosites, which are thought to be from the Moon's original crust, are represented in the fragments or 'clasts'.

Northwest Africa 482 meteorite

This is a slice of a lunar meteorite found in Northwest Africa in January 2000. Lunar specimens account for less than 1% of all known meteorites. Unlike the rocks collected and returned by the six Apollo missions, they provide a wider context of lunar geology as they come from all over the Moon's surface. The mineralogy and chemistry of this rock shows it is similar to rocks collected by the Apollo 15 mission.

Juice spacecraft

The Jupiter Icy Moons Explorer (Juice) spacecraft was launched in April 2023 and is due to arrive at Jupiter in July 2031. Its main objective is to study Jupiter and its moons Ganymede, Callisto and Europa. Juice's scientific instruments have been designed to provide a wealth of information on these moons, which form part of what is known as the Jovian system, including looking for evidence of habitable environments in their subsurface oceans.

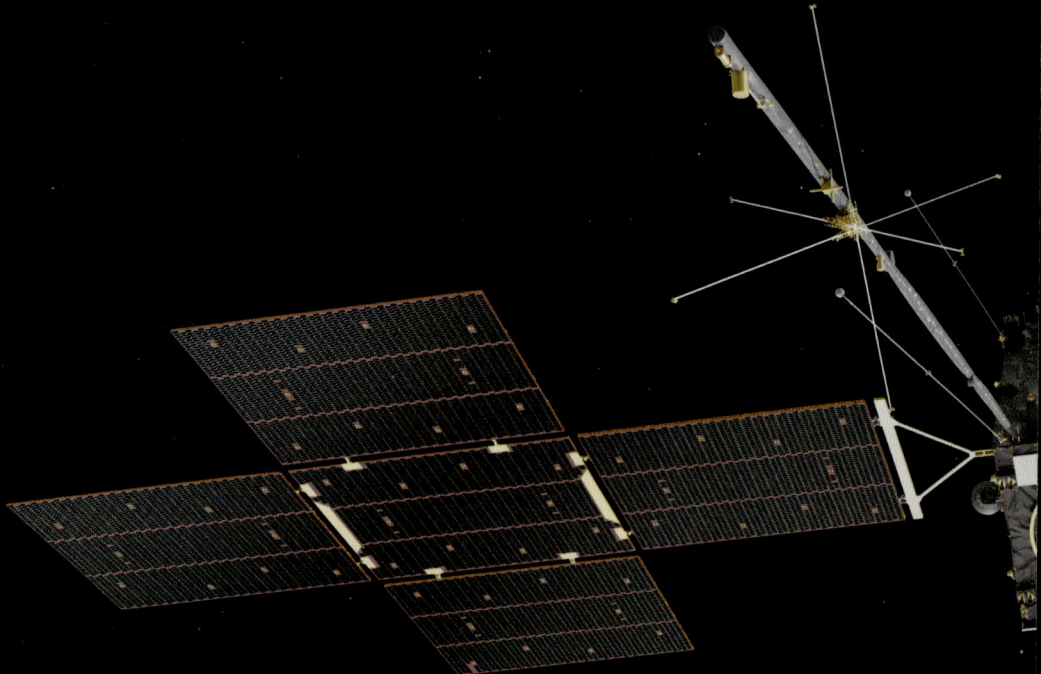

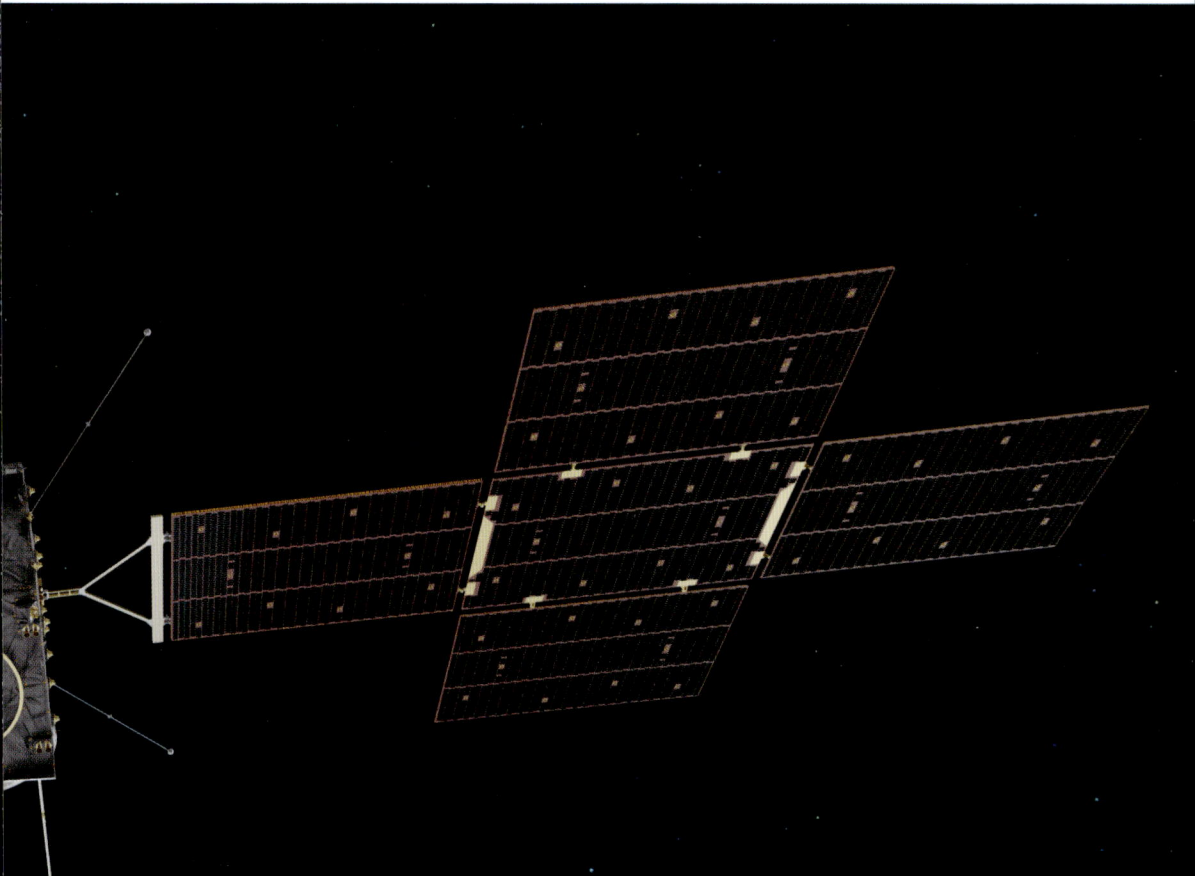

Magnetometer

A magnetometer is an instrument that detects and measures magnetic fields. The J-MAG magnetometer is a core instrument on the Juice mission, designed to measure in detail Jupiter's magnetosphere and to better understand the interior structure of Ganymede, Callisto and Europa. The J-MAG instrument is an update of the MAG instrument, designed by a team from Imperial College, London, which flew on the Cassini-Huygens mission.

Dragonfly lander

The Dragonfly lander is an innovative NASA space mission planned to launch in 2028 and arrive at Saturn's moon, Titan in the mid-2030s. The mission is based on a rotorcraft design that will be able to fly to, and then land at, many interesting locations on Titan's surface. Titan's environment is thought to be similar to that of the early Earth, and so could tell us about the conditions that led to life beginning on Earth.

In the depths of Earth's oceans, life thrives around hydrothermal vents. While most ecosystems are based on plants and microbes that produce food using sunlight, these communities are based on microbes that use Earth's own heat. This astonishing discovery opened up the possibility of life far away from sunlight, even in the dark oceans of distant worlds.

These creatures help us understand how life exists in the extreme pressures and temperatures of the deep oceans on Earth, an environment previously thought completely inhospitable for life. Knowing this, we can begin to see how life may exist in oceans of moons around Jupiter, Saturn and Neptune.

This hydrothermal vent chimney was seen in 2016 in the Mariana Trench region of the western Pacific Ocean. The vent fluid in the centre of the image appears like dark smoke due to the high levels of minerals and sulfides contained in the fluid. The chimney is crawling with shrimps and crabs.

Hydrothermal vent chimney and vent hood

In a world without sunlight, towering structures rise from the ocean floor, spewing hot, mineral-rich water. These fragments are pieces of these structures – a hydrothermal vent chimney and hood. The vents provide the conditions for communities of life to thrive, despite the extreme and unique environment. These hydrothermal vents from the East Pacific Rise are primarily composed of copper and iron sulphide minerals. Similar vents may be present in the sub-surface oceans on the icy moons of Jupiter and Saturn. These moons are being targeted by space agencies searching for potential habitats capable of supporting life.

Bathyaustriella thionipta
Glover, Taylor & Rowden, 2004
Loc: Kermadec Ridge, Macauley
Cone, New Zealand.
30° 13·04'S ; 178° 27· 112'W
480–504 m
Pres. J. D. Taylor, 2004
Specimens fixed in formalin
then transferred to 80% IMS.

Regis...
number...

VENT MOLLUSCS

Vent clam, *Bathyaustriella thionipta*

This deep-sea clam lives at depths of 480–500 metres on hydrothermal vents seeping from submarine volcanoes. In this inhospitable environment, the molluscs host bacteria in their gills that derive energy from both sulphides and methane. These chemicals compounds are found on other planets, and it is possible that extra-terrestrial life could use them as a nutrient source in a similar way.

Vent mussels, *Bathymolodius azoricus*

Life in the most extreme deep-sea environments on Earth has only been recorded relatively recently. *Bathymodiolus*, a genus of deep-sea mussels, was first discovered in 1977. The Atlantic species *Bathymodiolus azoricus* lives in huge numbers on hydrothermal vents, reaching densities of up to 10,000 individual mussels per square metre. Other species of *Bathymodiolus* are found in cold-seep environments. The ability to adapt to such different extremes might be useful for life on other planets or moons.

Scaly-foot snail

The scaly-foot snail, *Chrysomallon squamiferum*, was first discovered in 2001 at Kairei hydrothermal vent field, on the Central Indian Ridge (the north–south mid-oceanic ridge in the western Indian Ocean), and later found in four other vent fields. It lives at depths of between 2,400 and 2,800 metres (7,875–9,185 feet), on the edge of hydrothermal vents and black smokers that can reach temperatures of 300–400°C (572–752°F). The foot of this snail is unique among gastropods because it is covered in hundreds of iron-infused scales, which are fleshy in the centre and hard on the exterior. Research suggests that the primary function of these scales is the detoxification of sulphur metabolites, originating from symbiotic bacteria that the snail cultivates inside its cells for nutrition. Iron is a common element on Earth and in the rest of the Solar System and the galaxy. So the fact that this snail has incorporated iron in its shell and foot shows this is possible and so could be an evolutionary adaptation elsewhere, where biology has or is occurring in an iron-rich environment.

VENT WORMS

Tube worm, *Riftia pachyptila*

Giant tube worms live inside a tough tube, which they create themselves, anchored near hydrothermal vents. They have no digestive system – no mouth, stomach or anus. Instead, they rely on microbes living inside them that convert chemicals from deep-sea volcanic vents into food, allowing the worms to thrive in this extreme environment.i

Pompeii worm, *Alvinella pompejana*

The Pompeii worm is the most heat-tolerant animal known on Earth. Its tail end can survive in the scorching 80°C (176°F) waters around deep-sea hydrothermal vents – a deadly temperature for most animals – though it mostly lives in and prefers temperatures of 50°C (122°F). Remarkably, the worm's blood stays around 20–30°C (68–86°F), probably through a combination of the worm's unique blood and a strategy of shuttling cool water into its tube-like home. Up in space, on one of Saturn's moons, NASA have found evidence of gas plumes indicating the possible presence of vents in the moon's sub-surface oceans. Could similar life exist there?

Hoff crab, *Kiwa tyleri*

The Hoff crab is found in the cold Antarctic
waters of the Southern Ocean at depths of
2,500 metres (8,200 feet). Most crabs and
lobsters seize up and are unable to move
in extreme cold, so the Hoff crab lives on
hydrothermal vents, close to areas where
temperatures can reach 350°C (660°F). Like
many animals that live in dark, deep-sea
environments, it has no eyes. Its furry-looking
claws are covered in bristles that are home to
millions of bacteria, which produce most of the
nutrients the crab needs to survive.

VENT SHRIMPS

Alvinocaris markensis

These shrimps are both the scavengers and predators of Atlantic deep-sea vents. With no sunlight to guide them, instead of eyes they rely on other senses to scavenge food scraps and even hunt other small animals in the extreme darkness. The genus *Alvinocaris* is named after Alvin, the first research submersible to visit deep-sea hydrothermal vents in 1977.

Rimicaris hybisae

In the pitch-black depths of the Caribbean Sea, these shrimps 'see' with their backs. A special organ detects the faint heat of nearby hydrothermal vents, guiding them to food and safety. Shrimps found close to the vents farm bacteria on their claws, which provide them with most of their energy. Further away from the vents they survive by eating other crustaceans and molluscs.

Beyond

Sara Russell

Although we have a mere eight planets orbiting the Sun in our Solar System, there are countless more orbiting other stars in our galaxy. These are called exoplanets and their study is one of the most exciting fields of modern astronomy, especially as we search for planets that may look like our own habitable world.

Exoplanets can be detected using several techniques. The first identified exoplanets were discovered in the late 20th century using the radial velocity method. This technique uses the wobble of the parent star to show that there is a planet orbiting around it. Other techniques have also since been developed. Transit photometry involves the detection of exoplanets by observing the dimming of light emitted by a star as a planet passes in front of it. For this to be successful, the planetary system must be in exactly the right plane for us to observe it edge-on. A more recent technique is direct imaging of the planet, by blocking out the light from the parent star that would swamp any astronomical image. Finally, gravitational microlensing involves observations of the gravitational field of a star and planet because the planet can act as a lens, magnifying the starlight.

We can expect the field of exoplanets to continue to expand. Several new generation telescopes can observe exoplanets and build up a picture of their numbers and diversity. Missions like the Transiting Exoplanet Survey Satellite (TESS), which launched in 2018, will continue the work of the Kepler mission to expand our inventory of exoplanets, particularly around nearby stars. The James Webb Space Telescope (JWST), which was launched in 2021, can make detailed measurements of exoplanets, including their atmospheric composition. The next generation of telescopes, such as the Extremely Large Telescope (ELT), will provide unprecedented capabilities for direct imaging and atmospheric analysis of Earth-like exoplanets.

These techniques have all shown that there is a huge diversity of exoplanets. Many multi-planetary systems have been discovered, such as the seven planet TRAPPIST-1 system, and we are starting to learn something of their composition and atmospheric make up. Most of the first exoplanets to be discovered were 'hot Jupiters'. These are very large planets that orbit close to their parent star, which makes them easier to spot using radial velocity techniques. They can be so close that they complete an orbit within a matter of days and have challenged our understanding of how planets move in relation to their parent star. The most abundant currently observed planets are either Super Earths, planets a bit bigger than our own, or mini-Neptunes, that are a little smaller than Neptune. While the former can be either rocky or gas in composition, the mini-Neptunes are gas planets. Planets very close to their parent star may be covered in molten lava.

The study of exoplanets shows that planets can be incredibly varied in their size, composition and distance to their parent star. They also represent all ages, from newly forming planetary systems to ancient systems near the end of the star's lifetime, giving astronomers insight into the evolution of planets.

The abundance of exoplanets raises the exciting possibility that some of these planets may harbour life. Many searches focus on identifying terrestrial, Earth-sized planets that orbit within the 'habitable zone', that is, at a distance from their parent star where water may be in liquid form. Recent studies have even been able to detect the composition of the atmospheres of some exoplanets, using a technique called transmission spectroscopy, which uses the starlight that has passed through the thin atmosphere to determine its composition. Evidence of water, oxygen or methane, for example, are biosignatures that raise hopes of habitability.

Not only are these searches enabling us to learn more and more about how planets form and evolve, observations of Earth-like planets are the places most likely to host a habitable environment like our own and may be the home of complex life beyond our Solar System.

OVERLEAF
The Extremely Large Telescope (ELT) is a new-generation ground-based telescope that has been built in the high Atacama desert in Chile. The ELT has a massive 39 metre (130 feet) diameter main mirror, and this will enable us to directly image exoplanets. It will even be able to detect atmospheres around exoplanets, and maybe find biomarkers suggesting that they are habited.

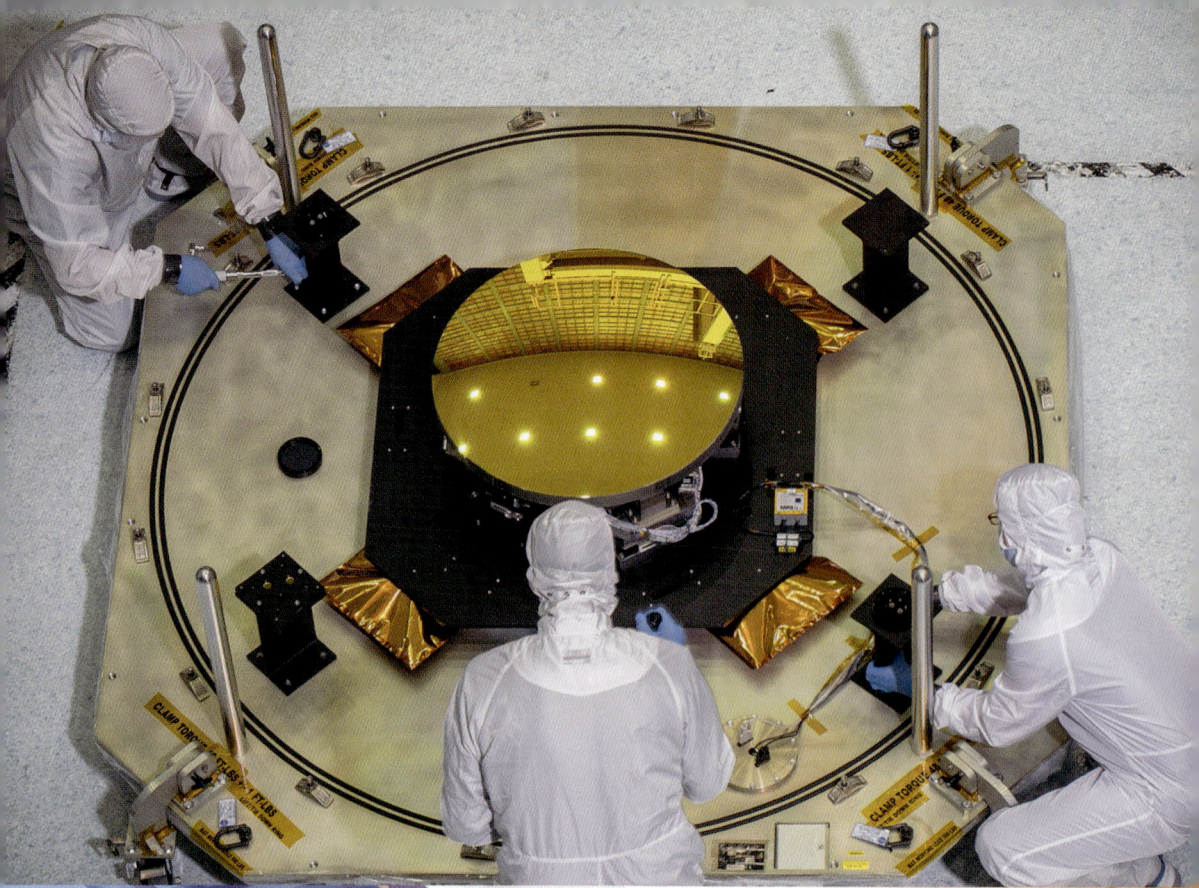

Mirror on James Webb telescope

The James Webb telescope is a space-based observatory capable of imaging Earth-like planets orbiting other stars. The mirrors in the telescope are a key part of the story as they focus and reflect light that allow measurements to be made of far away exoplanets. Light from a star with a planet orbiting it shines through the atmosphere of an exoplanet and is received by the telescope. This light can tell us about the chemistry of these world's atmospheres.

LaserSETI device

From the hills of California to the volcanic peaks of Hawaii, a team of scientists is using cutting-edge technology to search for evidence of intelligent life beyond our planet. The LaserSETI project, with its observatories strategically placed across the globe, is monitoring the entire sky for any unusual flashes of light that could indicate 'techno-signatures' – the traces of technological activity used by a civilisation. Unlike other SETI efforts that listen for radio signals, LaserSETI is designed to detect bursts of light so brief and focused that they would be nearly impossible to produce naturally. These could be the telltale signs of powerful lasers used for communication or even interstellar travel. The key to LaserSETI's success lies in its innovative spectroscopy technique. Rather than peering through a narrow slit, the instrument's cameras can analyze the full spectrum of light from countless stars simultaneously, searching for any anomalies.

Cold Bokkeveld meteorite

This 'Mighei-type' carbonaceous chondrite was observed to fall in Cape Province, South Africa on 13 October 1838. Carbonaceous chondrite meteorites are rich in clay minerals, which formed through the interaction of water and rock on the parent asteroid. They also contain organics, such as amino acids and hydrocarbons, and are believed to be a major source of the organic material on Earth.

Pre-solar grains and Murchison residue bottle

Pre-solar grains formed before the Solar System came together 4.567 billion years ago. They are composed of tiny solid grains that each originate in a single star that was an ancestor to our own Sun, and so can give us a glimpse into processes happening in other stars. When some meteorites are dissolved in very strong acids, the residue remaining can be enriched in these precious crystals. This is a vial of such a residue, containing precious pre-solar silicon carbide along with diamond and organic material that are likely to be a mixture pre-solar and early Solar System in origin.

Picture credits

p.7 ©Alex Pérez – ALMA; p.8/9, 49, 50, 52, 53 ©NASA/JPL-Caltech; p.26 ©Frances Westall; p.34, 35 ©JAXA; p.38 ©NASA/JPL, Public domain, via Wikimedia Commons; p.39 ©NASA Goddard's Scientific Visualization Studio; p.40, 41 ©NASA/Keegan Barber; p.42 ©NASA/Erika Blumenfeld and Joseph Aebersold; p.45 ©NASA/USGS; p.47 ©NASA/JPL-Caltech/MSSS; p.48 ©ESA; p.51 ©NASA/JPL-Caltech/ASU; p.55 ©ESA/ATG medialab; p.57 (top) ©M. de la Nougerede, UCL/MSSL; (bottom) ©UCL MSSL; p.65 ©NASA/JPL-Caltech/Space Science Institute/University of Arizona/DLR; p.66 ©Image data: NASA/JPL-Caltech/SwRI/MSSS Image processing: Kevin M. Gill CC BY 3.0; p.70/71 ©ESA (acknowledgement: ATG Medialab); p.72 ©Imperial College London; p.73 ©NASA APL, Public domain, via Wikimedia Commons; p.74 ©NOAA; p.88/89 ©European Southern Observatory; p.90 (top) ©NASA/Chris Gunn; (bottom) ©NASA/Drew Noel; p.93 ©Eliot Gillum;

Unless otherwise stated images copyright of Natural History Museum, London.

Acknowledgements

This book was written to accompany the exhibition *Space: could life exist beyond Earth?* held at the Natural History Museum, London, 2025. We'd like to thank the following for their support during the compilation: Karen Bell, Callum Slade, Emma Sherlock, and Tom White (from the Natural History Museum), and Keyron Hickman-Lewis (from Birkbeck, University of London) for imparting their knowledge on specimens; Paolo Cocco, Miranda Hine, Sinead Marron and Amy Pollack, exhibition writers, for their content input; and Lucie Goodayle, photographer for expert specimen photography.